G. B Hetley

The native flowers of New Zealand

Illustrated in colours in the best style of modern chromo-litho art

G. B Hetley

The native flowers of New Zealand
Illustrated in colours in the best style of modern chromo-litho art

ISBN/EAN: 9783741176135

Manufactured in Europe, USA, Canada, Australia, Japa

Cover: Foto ©Klaus-Uwe Gerhardt /pixelio.de

Manufactured and distributed by brebook publishing software
(www.brebook.com)

G. B Hetley

The native flowers of New Zealand

THE

NATIVE FLOWERS

OF

NEW ZEALAND

ILLUSTRATED IN COLOURS

IN THE BEST STYLE OF MODERN CHROMO-LITHO ART, FROM DRAWINGS
COLOURED TO NATURE

BY

Mrs CHARLES HETLEY

LONDON:

SAMPSON LOW, MARSTON, SEARLE, AND RIVINGTON (LIMITED)

ST. DUNSTAN'S HOUSE

FETTER LANE, FLEET STREET, E.C.

1888

Dedicated
by
Special Permission
to
Her Most Gracious Majesty,
Victoria,
Queen of Great Britain & Ireland
and
Empress of India.

PREFACE.

The intention in publishing this book is to make known—quite as much in New Zealand as elsewhere—some of the lovely wild flowers we possess, many of which have never been painted, and some others only from dried specimens, as in Hooker's Flora. This is now out of print, and no coloured illustrations of the New Zealand Flora have been published since.

I have often heard it stated that we had no flowers in New Zealand, or very few, such as the Ratas, the red and yellow Kowhai, the Clematis, the Tea Tree, and a few others. When travelling about the country collecting flowers, I was much astonished to find how numerous and varied they were. Botanists no doubt knew of them, and had them carefully dried; but they were pressed out of shape and colour, labelled, and put away in cases out of sight, (unless required for reference), and each one taken from its fellows and put with others of the same genera from other parts of the world. It was on hearing a lecture from Mr. Cheeseman, in Auckland, after his return from an expedition to the mountains about Nelson, and to Arthur's Pass and the Otira Gorge, Canterbury, when he showed us dried specimens of beautiful flowers of different colours, and described how lovely they looked growing in their native state, that I first thought, what a pity it was they were not painted. Then again, several years later, 1884, when I had begun to paint a few myself, I applied to Mr. Cheeseman, at the Museum in Auckland, for the names of some of them. He showed me three of the same genus, and said that this was yellow, this one was white, and this purple. They were all reduced to a dirty looking brown colour. I again said, "What a pity they are not painted." A gentleman present said, "Why don't you do it?" I said, "I would if Government would help me." This the Government did. The Honourable E. Richardson, the Minister of Public Works, kindly gave me passes on the railways, and the Minister of Education, Sir Robert Stout, took copies of my book for public schools and libraries. Mr. Mills Managing Director of the Union Company, kindly gave me passes on the steamers, and later on, the N. Z. Shipping Company offered me a saloon passage to England for the price of a second class, on account of my making known the New Zealand flowers, but I was unfortunately unable to avail myself of the offer.

It was a serious undertaking, for I had to travel by sea and by land, coaching over rough and dangerous roads, and at great expense, risk, and fatigue. But it was a labour of love. Every new flower was a delight and wonder; and the scenery, which I might otherwise never have seen, and the delightful excursions with kind friends to help to get flowers for "The Book," was enough to repay all my fatigue.

My first excursions were near Auckland, at Wai-wera (hot water), where the stream of hot water literally pours out of the cliff into delightful baths, with its comfortable hotel and lovely scenery, with its two rivers, where you go up in a boat between wooded mountains and under trees with ferns hanging from the branches, or with bushes of white Tea-tree (*Leptospermum scoparium*) their boughs covered with little white flowers, looking like long plumes bending over almost into the water, reflected distinctly on the smooth surface. From the delightful beach of hard sand, may be seen bold cliffs with huge trees of *Metrosideros tomentosa* (plate 29), clinging to the face, and on the hill above, amongst the "bush" of different kinds of trees and creepers of every shade of green ; you may see in the Spring patches of bright yellow Kowhai (*Sophora tetrantha*) Clematis, white Pukapuka, or New Zealand Lilac (*Brachyglottis repanda*), with its large leaves—silvery white underneath—and heads of small cream flowers. Then across the valley on the wooded hill opposite, may be seen large trees 50 feet high, a mass of bright red, these are Rata trees (*Metrosideros robusta*). In the woods, festooned from tree to tree, there is the bramble or Bush Lawyer (*Rubus Australis*), with long sprays of white flowers, and with it the bunches of scarlet berries of the Supple Jack (*Rhipogonum scandens*), which forms such a thick twisted mass in the uncut "bush," that you can only get through it with difficulty, and may often be tripped up or half hanged by it. Then all around on the stems of the trees, and hanging from the branches, and on the ground, are lovely ferns of all shapes and sizes. All the flowers I have described, and many others, I have seen in blossom at the same time at Wai-wera in Spring. These are common ones, and may be found at Whangarei, Waikomiti, the Thames, and other places near Auckland. At the Thames there are many additional ones, some of which Mr. Adams took much trouble in procuring for me, and others I saw for the first time in a ride up the Ranges, for instance, the lovely white Rata (*Metrosideros albiflora*) (plate 18), hanging down a bank and climbing the branches of the trees in masses of white feathery balls, the tall forest tree of *Quintinia serrata* (plate 33), in full blossom, the pretty little trees of *Phebalium nudum* (plate 32), and the graceful *Senecio myrianthos*.

My next collecting ground was beautiful Taranaki, "The Garden of New Zealand," as it is called, my old home, where we went through the war and had houses burnt, and sheep, cattle, and horses carried off by the natives. It is a beautiful region with its grand peak of Mount Egmont, its mountain streams running over rocks and stones between high cliffs and wooded mountains, with ferns, especially the *Hymenophyllum* tribe, everywhere, *H. demissum*, *H. polyanthos*, and *Trichomanes reniforme*, carpet the ground in many places, as may be seen in the Forster case in Kew Gardens. In walking up the path, or rather water-course, up the Ranges, 5,000 feet high, I could not resist stopping continually to gather *H. pulcherrimum*, *H. acruginosum*, *Todea superba*, and the most beautiful mosses with large fern-like leaves, growing on the tree trunks and rocks. When about 4,000 feet high we emerged from the "bush." The view was supurb. It seemed as if for miles and miles there was nothing but trees and the sea beyond them. The town of New Plymouth lay far away in the distance, but we could not see it, it was hidden by the smoke of the burning "bush." The beautiful forest with its flowers and ferns is fast disappearing before the tide of cultivation, and many will only be known by their dried and shrivelled up remains. The short scrub where we were, was greatly composed of *Senecio cleagnifolius*, or "*brown lacks*" (plate 15), and a very curious kind of spear grass. I could not get any further, but others of the party went to the top of the ridge, another 1,000 feet, and they kindly brought me several lovely little Alpine flowers, *Forstera Bidwillii, Euphrasia Monroi, Gnaphalium bellidioides*, and others. At the foot of the Ranges the small Snow-trees (*Carpodetus serratus*), so called from the quantities of little white flowers making it appear as if there had been a fall of snow, were in full bloom, the flowers all down the stems, with shining leaves on each side, forming the most lovely wreaths. After a twelve hours' journey by train to Palmerston, I started at six the following morning on my overland journey by coach to Wellington, going through the famed Manawatu Gorge, sleeping one night on the way, then starting at 4 o'clock a.m., and on through miles and miles of forest with some good bits of mountain scenery to Masterton, then by train to Wellington, zig-zag up the Rimutaka Mountain, with the great engines (four I think), puffing and snorting as if they hardly could get up. We looked down on the beautiful scenery, the trees red with Tetoki berries, passed the place where a train, engines and all, was blown down the hill by the wind, and then went through the tunnel and down the other side to the Hutt Valley. After going by sea to Nelson, through Queen Charlotte's Sound and the French Pass, we went by coach to a station called Lake Station, belonging to Mr. John Kerr, whose family were most kind in helping me. The roads in New Zealand are very narrow, with only just room for the coach, and no wall or anything to prevent one going over the precipice. Once as we were at the top of a mountain range, and had gone round a sharp curve, one of the traces broke. Some evil disposed men had wantonly set fire to the forest all along the road, burning whole sides of mountains and destroying some of the most magnificent scenery, we were several days in going through, it was still smoking and occasionally blazing up, and there was the risk of burnt trees falling on us. After leaving the coach we were driven by buggy twelve miles to the station, and had to pass over a burnt wooden bridge, which it was hoped would not give way. We went in a boat on a lovely lake, Roto-iti (or Little Lake), with high mountains round three sides of it ; we landed on a point and walked up to a waterfall where I got the *Loranthus Colensoi* (plate 30). *Ourissia macrophylla*, and some others. At the edge of the lake there were bushes with lovely berries of different colours, and such large white snow berries (*Gaultheria antipoda*).

Wahlenbergia saxicola, and *Gentiana saxosa* (plate 24), were in great abundance, also some of the Veronicas, forming great masses of white and lilac.

It was now February; the season was too advanced, and most of the Spring flowers had disappeared ; it was a great pity, but I had been detained both in Wellington and Nelson. I was anxious to hurry on so as not to lose those in Arthur's Pass. We drove again the twelve miles, and passed the burnt bridge a second time in safety, arriving at the little inn amongst the mountains, where we joined the coach. The view was most beautiful of the river winding down the valley many hundred feet below. The sides of the road were lined with *Lomaria vulcanica*, *Aspidium vestitum*, *Hypolepis distans*, *Adiantums*, and other ferns. There were long bunches of blackberries and glossy black mako-mako berries, which we saw all the way to Hokitika. We went all along the banks of the Buller River, losing it now and then and coming back to it, seeing some of the most beautiful scenery in New Zealand, and by a road not much travelled, unfortunately we passed some of the finest parts in the dark. The drivers of the coaches on this road are brothers. They are most careful, and it is very necessary that they should be so. Your heart is in your mouth most of the way. At one place in particular, the road is built *outside* the cliff, and supported on piles, which are inserted somehow into the rock. The cliff rises perpendicularly above you, and there is only just room for the coach to pass round without touching, and there is hardly an inch to spare on the outside edge which has no wall or fence. If one of the horses shied or fell, coach and all would go over into the river, which rushes along two hundred feet below, and we saw all this from a turn in the road before we came to it, which made it worse. I kept my face turned to the cliff, but my niece, who was with me and had a stronger head, kept calling my attention to the magnificent scenery. We both drew a long breath when it was over, and were truly thankful to be safely through ; yet the coach goes every day with the same driver. The chief danger I believe, lies in some of the wooden piles becoming decayed, how the road was ever made is marvellous. We thought very little of the famous pass in the Otira Gorge after this, though this is considered very bad. We travelled some distance in the dark, and so I missed *Senecio Hectori*, but Mr. Buchanan kindly supplied me with it afterwards (plate 20).

We arrived about 9 p.m. at a small inn, and started at 6 a.m. next day through a cold mist from the river, which cleared away afterwards, and it became very hot. We were soon passing through country with gold-digging shafts, water-races, and other traces of mining. We crossed several rivers, sometimes bumping over the big boulders and struggling through the rushing water, others by ferry, and at one, the Teremakau, we left the coach and entered a kind of wooden box, hung on a rope, which was wound up by a small steam engine on the other side. We slid down one side and up the other. It was not an unpleasant but a very curious sensation to find oneself suspended from one to two hundred feet above a broad rapid river.

On one occasion the rope which pulled the cage on one side broke, and the passengers were dangling by the other over the roaring torrent until assistance came.

When we landed at the opposite side we entered a bush tramcar drawn by one horse, a very primitive arrangement, on rails of wood. After travelling some distance we got into another coach which was full of Chinamen, and reached Greymouth in the evening, very glad to get once more into a comfortable hotel. The next morning we started by tram again. We had it all to ourselves, and enjoyed the quiet, smooth, plodding along through a narrow lane in the bush, always the same avenue stretching away in the distance, whether we looked behind or ahead, there were tall forest trees, and masses of creepers, ferns, mosses, lichens, and flowers, which I longed to gather, but we could not stop. Then we went on by coach again, driven by a young gentleman, son of a military man, who entertained us by his experiences till we reached Hokitika, where we stayed four days waiting for the coach to take us across the mountains to Christchurch. It rained of course, most of the time ; it always does at Hokitika. But one clear lovely morning we had a splended view of Mount Cook, and the Southern Alps covered with snow in the far distance, standing out hard and clear against the sky, touched with pink and yellow by the rising sun.

We thought we had secured the lower box seats for the journey to the Otira Gorge, but found ourselves mistaken, and had to be content with the upper ones on the top of the coach, high up, with our feet even above the driver's head, and the horses in the depths below, with nothing to prevent us from slipping off. After one long very steep hill, down which we went full gallop, I gave in and went inside, until a lady down below kindly exchanged seats with me. The first part of the journey was comparatively easy, with a good road between wooded mountains, already becoming red with the Rata flowers (*Metrosideros lucida*), and over streams, some not bridged, down we went at full speed, and up the other side, and woe betide us if the springs broke. About 8 a.m., we came to a small mining village in the midst of the forest, where we had breakfast. We drove through beautiful scenery till we reached the Taipo, or Devil River, where we all had to leave the coach and cross by a long narrow swinging bridge, which was rather a trial, as it oscillated so much. We got into the coach again on the other side, and to judge by the way the luggage was mixed up in the inside, it was well we had some other means of crossing.

About 2 p.m., we arrived at the Otira Gorge, where there is a small inn by the side of the river, which is very broad there, or rather the bed of it is, as there are two channels, the rest is all stones. We dined here, and the other people went on by the coach, but my niece and I stayed till the following one, four days later, in order to obtain the flowers on Arthur's Pass, which is an excellent place for Alpine plants. I prospected the river bed and side that afternoon, and found plenty of Veronica Lyallii, a small plant with delicate pale lilac flowers, *Raolias Rhabdothamnus* with bright crimson leaves, trees of *Olearia* in bloom, and the ground strewn with *Lomaria alpina* ferns. Just before crossing the river there were magnificent plants of *Todea superba*, of which our landlord kindly got me a large quantity carefully packed up, but alas, they had to be left behind as the coach was too full. The following day he went with us to Arthur's Pass, my niece and I riding in turns, and we very much enjoyed our expedition. We saw the celebrated gorge to great advantage in light and shade, which is generally wanting, as the coach passes through in the afternoon when the sun has left it. It was shining brightly in the morning when we went, and the effect was most beautiful, especially from the bridge across the gorge, where we could look up and down. Near the top the Hoheria or Lace-bark trees (*Plagianthus Lyallii*) (plate 34), were in full blossom with bunches of cherry-like flowers, and close alongside of them were trees of crimson Rata (*Metrosideros Lucida*), making a lovely contrast; also bushes of Veronica, a mass of white flowers (*Olearia ilicifolia*) (plate 21), and many others. We descended into a valley and then mounted up to Arthur's Pass. I was very sorry to find most of the flowers were over, but I could see how many more I could get if I came earlier in the season on another occasion. I was delighted to find a small piece of New Zealand Edelweiss (*Gnaphalium grandiceps*) (plate 31), the beautiful *Celmisia Monroi* (plate 5), the plants of which were in such profusion that the whole Pass must have been studded earlier in the year with its large starry white flowers. There were *Veronicas Senecios*, *Celmisias*, of different kinds, the little *Utricularia*, lovely white *Gentians* (plate 24), the long-spike of spear-grass, and many others. In the valley we had passed through by the side of the stream, there were numerous plants of Ranunculus Lyallii, or Mountain Lily, as it is called. We returned in the evening through the gorge, with baskets full to overflowing with flowers, only a few of which I could paint at the time, others we packed in tins and took with us to Christchurch. Some we pressed, as well as the ferns, and were kept busy until the coach arrived by which we went on. The lovely *Veronica Lyallii* was growing by the side of the road. On the other side of Arthur's Pass we had a glimpse of a glacier. By-and-bye we came to a narrow road, made of loose boulder stones, with a steep slope down into the river far below, which had often before given way. We felt anxious for those on the top of the coach, some of whom had their feet dangling over the side when it swung about, sometimes nearly over the edge. Further on we came to a pass down the side of a mountain. The gentlemen walked down a short cut, but we had to remain in the coach; the road was *very* narrow, and we went full gallop in and out, round the points and rocks which we nearly grazed, and once when we *jumped* a water-course we thought the coach was really over, it seemed to sway right over the precipice. But one lady stayed on the top all the way. After crossing sixteen times, the last time in the dark, the Waimakariri river, the bed of which is composed of round boulders, sometimes a quarter of a mile wide, the latter part when it was quite dark, we at last reached the Bealey Hotel where the coach from Christchurch had arrived, also full of people. The next day, after an early start, and much grumbling about boxes, &c., having to be left behind, we reached Springfield, and went on by train, arriving in the evening at Christchurch. Here I stayed six weeks painting; first the flowers I had brought with me in tins, and then some out of the native garden (the best in the colony), in the beautiful Botanical Gardens, where the Armstrongs, father and son, have cultivated the indigenous flora with great success, collecting the plants from the mountains, and also from Stewart's and the Chatham Islands. They were very kind in giving me those I wanted, and also packets of seeds which I brought over to my friends in England. They have many orders, but it seems very difficult to acclimatize the New Zealand flowers. It was intensely hot when we arrived, but soon after there was a fall of snow on the mountains, and I had to give up going to Lake Wakatipu, for which I was very sorry. We went for two days to Dunedin and back, and saw Mr. Buchanan, who, when he was the Government botanist, drew for the Transactions of the New Zealand Institute all the newly discovered plants, but he has now retired, a martyr to rheumatism, the usual result of exposure to our climate. He was much pleased that the flowers should be painted and made known before they were lost by burnings and the cultivation of the ground. He asked how he could help me, and kindly painted for me the two *Senecios* (plate 20), which I could not get, and gave me the paragraph published in my prospectus. I went back to Wellington, and overland to Taranaki by another route, by train and coach along the sandy beach for six hours, then by train again from Foxton, and steamer to Auckland, seeing some more beautiful scenery, and having some adventures, which, as my account is, I fear, already too long, I must not relate.

My sketches created some surprise in Auckland, as even people who were born in New Zealand had no idea we had such lovely flowers. I left shortly after for England in order to have my plates well done. On seeing some works similar to mine, published by Messrs. Sampson Low, Marston, Searle, and Rivington, beautifully produced, I determined to give them mine to do, and I have great reason to be pleased that I did so. I have to thank them for their great kindness and consideration, without which this book would never have appeared.

I am also much obliged to Messrs. Leighton Brothers for the way in which they have executed the plates, they have rendered my paintings exactly, except in a few cases. The proofs were sent to me for correction in Madeira, and the Portuguese Government detained the parcel in Lisbon two months. It was only from agitating through the Postmaster-General in England that I received them at all. This caused the long detention of Parts II. and III.; the plates were printed without my correction, consequently the colouring in plate 26 is not quite correct, and plates 18 and 25 are not so good as they should be.

Professor Kirk, Conservator of Forests, &c., and Mr. Cheeseman, Curator of the Auckland Museum, have kindly aided me throughout. My brother, Henry S. M'Kellar, Secretary and Inspector of Customs, New Zealand, and my brother-in-law, Dr. Hetley, of Norbury Lodge, Upper Norwood, have kindly assisted me greatly. I have also to thank Sir J. Dalton Hooker, Mr. Thistleton Dyer, Director of Kew Gardens, Mr. Morris, Assistant Director, Professor Oliver, and Mr. Baker, under whom I worked at the Herbarium at Kew. Also Mr. Rolfe, who kindly showed me how to make the dissections, found the books I required, and helped me in every way he could.

I enjoyed my stay at Kew writing descriptions, doing dissections, and obtaining information from the numerous books, many of which are not to be obtained elsewhere. I am sorry it is over and that my work is done for the present. I should be glad to continue it. I have many more flowers than I was able to put in these three parts ; some lovely and interesting ones I was sorry to leave out, and I could paint many more after my return to New Zealand, and make this work a more complete one, if I received sufficient encouragement to do so.

I was asked to put in dissections of the flowers, and, not having sufficient knowledge myself, I have traced all I could find from different books at Kew, and have named the authority from which I took them.

The others I have tried to do from some dried specimens I had, I am afraid they are of no real value botanically, but they may be of some assistance to amateurs. I trust my kind subscribers will excuse their faults and accept them for what they are worth. I feel how very imperfectly I have represented the flowers, but if I should be able to continue the work, the additional experience and the knowledge I have obtained at Kew, will, I trust, enable me to make my drawings of more botanical value and interest to the friends who have so kindly assisted me in the production of this work.

G. B. HETLEY.

INDEX.

Clematis indivisa, Willd.

Common large-flowered Clematis.

———————

THIS beautiful climber, one of the finest of the genus, is found throughout the whole of the lowland districts of New Zealand. It is thus a familiar plant to the colonist; and its pure white flowers, often produced in immense abundance, are welcomed as the harbinger of spring. They usually appear in August; but the maximum of bloom is not attained until the end of September. The curious feathery, tassel-like fruits are ripe about the end of the year. Considering the beauty of the plant, it is singular that it is seldom cultivated; but this is perhaps explained by the fact that it is so generally distributed as to be easily obtained in a wild state.

CLEMATIS INDIVISA. WILLD.

Olearia semidentata, Decaisne.

———◦•◦—— ·

A very beautiful and graceful shrubby plant, a native of the Chatham Islands, which have several remarkable species not found in New Zealand proper, some of which are figured elsewhere in this work. Cultivated specimens of *O. semidentata* have the flowers of a deeper purple than in the wild state, in which it is not uncommon to see the ray-florets almost white, tipped with pale purple at the extremities. It is a plant well worthy of general cultivation in gardens. It flowers in Spring.

Epacris microphylla, Br.

———◦◦◦———

THIS well known Australian plant is only known in a single locality in New Zealand, on the southern shore of the Manukau Harbour, and may possibly be introduced and not a true native. It has a pretty and graceful mode of growth, harmonizing very well with the *Leptospermum* and *Pomaderris* scrub with which it is usually associated. It flowers in November.

EPACRIS MICROPHYLLA,

PLATE 3.

Senecio perdicioides, Hook, fil.

———◦•◦———

A compact close-growing little shrub, 6 — 12 ft. in height, apparently confined to a very limited district near the East Cape (North Island) where it was first collected, over a hundred years ago, by Banks and Solander, the distinguished naturalists who accompanied Cook in his first voyage to New Zealand. It is not closely allied to any of the other species native to the country, and can be easily recognized by its small oblong waved and crenate leaves, nearly smooth habit, and rather small flower heads. It flowers in December.

Celmisia Monroi, Hook, fil.

—————◦◦◦————

One of the most beautiful species of a very beautiful genus. The numerous stiff, pointed, furrowed and wrinkled leaves, and the large Aster-like flowers, which are placed on stalks covered with shaggy white wool, give the plant a very singular and remarkable appearance. It is a native of the mountains of Canterbury and Marlborough, usually growing at altitudes of from 2500 to 4000 ft. It is particularly abundant on Arthur's Pass, on the coach road from Christchurch to Hokitika, and is there one of the chief ornaments of a varied and handsome sub-alpine vegetation. It flowers in January and February.

Professor Oliver says :—

"The plant figured differs from the typical *Celmisia Monroi* in its much larger leaves, and heads nearly or quite destitute of cotton."

CELMISIA MONROI.

MOUNTAIN RATA OR IRONBARK.

Metrosideros lucida, Menzies.

———◦◦◦———

A handsome tree, 30 to 50 ft. in height, not uncommon in hilly situations from the Thames river southwards. It is particularly abundant in the sounds on the South West Coast of Otago, and presents a magnificent sight when covered with its bright scarlet blossoms, which are often so abundantly produced as to clothe the whole forest with colour for miles together. The flowers are smaller than those of the true Rata *(M. robusta)* or Pohutukawa *(M. tomantosa)* but are very similar in structure. It flowers in January and February.

Pimelea longifolia, Banks and Sol.

A most charmingly handsome little shrub, seldom more than 6 ft. high.   In mode of growth, foliage, and inflorescence, it closely resembles some of the Daphnes now so frequently cultivated in gardens.   It is found throughout the North Island, and in the northern parts of the South Island.   It flowers in March.

PIMELEA LONGIFOLIA.

PLATE 7.

Native Name: — N I K A U.

Areca Sapida,

NIKAU PALM.

———◦•◦———

THIS is a very beautiful Palm, growing sometimes to a great height. The lovely lilac flower in its different stages from when it is packed into its sheath until the stage in which it is represented in plate 8, when it is commonly called a "hand," and afterwards when the same "hand" is covered with bright scarlet berries makes it a very beautiful object in the forest. The young infloresence is eaten and has an agreeable nut like flavour. It has simple ringed trunks, flowers very numerous and crowded, males and females mixed. It grows abundantly in the North Island and part of the Middle Island. It flowers in March.

ARECA SAPIDA.

PLATE 8

Native Name : — K O H E · K O H E.

Dysoxylum spectabile, Hook, fil.

A large tree, forty to fifty feet high or more, with spreading branches and large dark green and glossy pinnate leaves. The flowers are waxy white, and are remarkable for being placed in long pendulous panicles springing from the branches considerably below the leaves. The leaves are very bitter; and an infusion of them is said to possess valuable tonic properties. It is confined to the North Island. It flowers in July and August.

KOHE-KOHE

DYSOXYLUM SPECTABILE.

Geranium Traversii, Hook, fil.

A beautiful plant, confined to the Chatham Islands. It is allied to the world-wide G. dissectum but is easily distinguished by the larger size, stiffer leaves, covered with almost hoary down, and by the much larger flowers. It is well worthy of more general cultivation as a garden plant. It flowers in January.

GERANIUM TRAVERSII.

MOUNTAIN LILY.

Ranunculus Lyallii, Hook, fil.

THE finest known species of Buttercup, and one of the most magnificent plants native to New Zealand. The leaves are perfectly round, with the stalk affixed to the centre. They are often from one to two feet in diameter, and are concave on the upper surface, so as to somewhat resemble large soup-plates. The stem is often over four feet in height, and is usually much branched. The flowers are very numerous, pure white, and from two to three inches in diameter. It is confined to the South Island, and is purely an Alpine plant, being most abundant at an altitude of three thousand feet. In some of the valleys in the central portion of the Canterbury Alps it is exceedingly plentiful, and during the flowering season the slopes of the mountains are often whitened from the abundance of its blossoms. It flowers in January.

MOUNTAIN LILY

RANUNCULUS LYALLII.

Loranthus Adamsii.

———⁃•▴•⁃———

This is a new species of Loranthus which has never been figured before. It was discovered by Mr. James Adams of the Thames Gold Fields, who kindly obtained this specimen for me. It is a small bush two to three feet in height. The flowers are rather large, one and a half to two inches long, and of a very peculiar shape. It flowers in September and October.

SENECIO HUNTII, *F. Mueller*.

This fine species is confined to the Chatham Islands, where it is not uncommon, attaining a height of 25 ft. The branches are often bare of leaves except at the extremities, so that the habit of the plant is not so handsome as in others of the shrubby species. The fine, large heads of yellow flowers are, however, very ornamental, and the plant has consequently gained itself a place in many colonial gardens. (*Cheeseman*). It flowers in February.

SENECIO HUNTII

ANTHERICUM HOOKERI, OR CHRYSOBACTRON HOOKERI, *Hook. fil.*

THE first species of the present genus (C. Rossi) was detected by Dr. Hooker in Lord Auckland Islands. It was named Chrysobactron "in allusion to the magnificent racemes of golden flowers (staff of gold) which that species bears." C. Hookeri is a less showy species. (*Bot. Mag.*) It grows on Mount Egmont (Taranaki) and on the Canterbury Plains, Middle Island. It flowers in December.

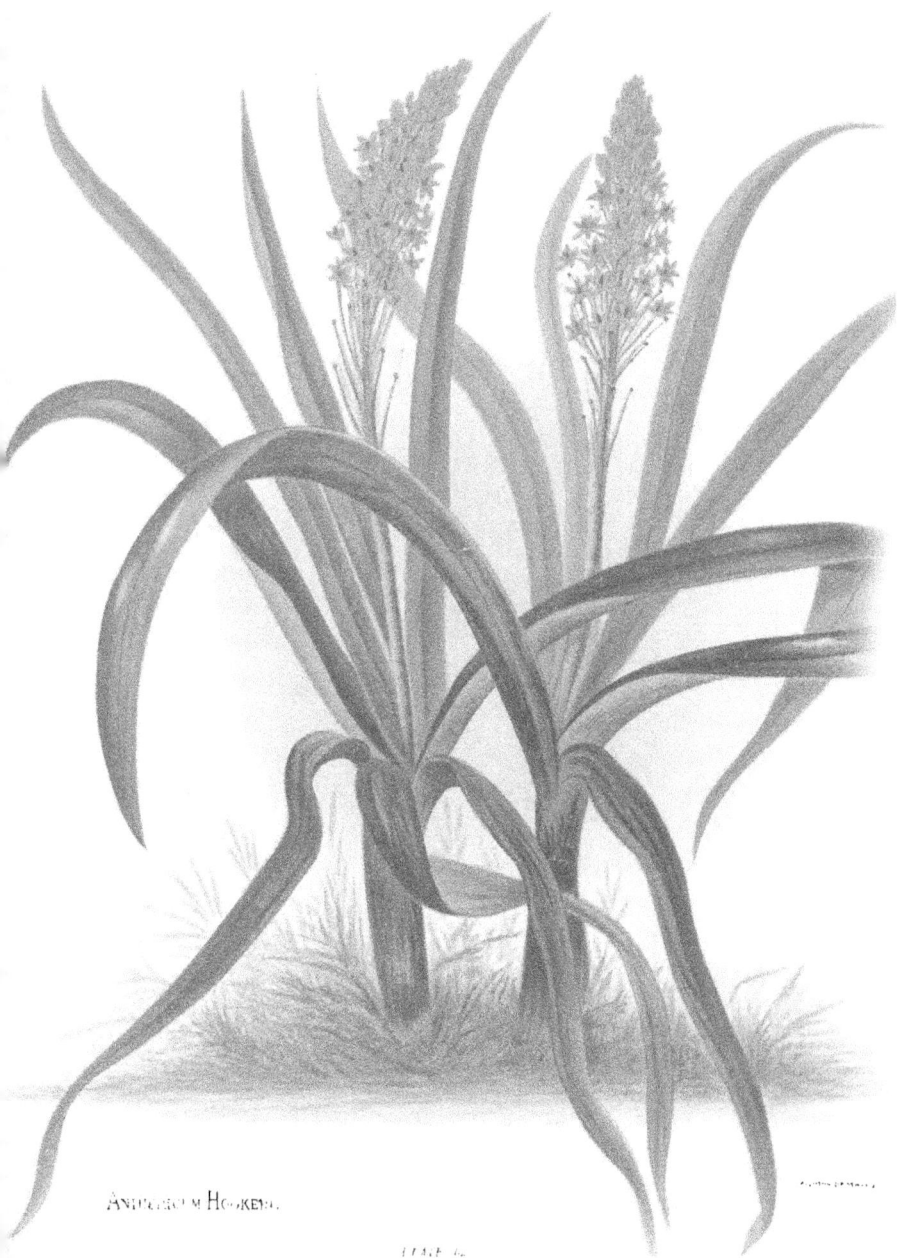

ANTHERICUM HOOKERI.

PLATE 1.

SENECIO (BRACHYGLOTTIS) ELEAGNIFOLIUS,

Hook. fil.

Brown Backs.

A SMALL robust shrub, 6 to 8 ft. high. Leaves extremely thick, hard, and leathery; upper surface glossy, smooth, and of a bright green, with white trellis-like markings; densely covered beneath with brown tomentum or hairs, hence its name; head without a ray; flowers all tubular. *Hook. fil.* It grows in the Northern and Middle Islands. The specimen represented came from Mount Egmont, on the Ranges 4,000 ft. high. Flowers in January.

METROSIDEROS FLORIDA, *Hook. fil.*

Native Name:—KAHIKAHIKA.

THIS is found throughout the Islands in forests, and has a most brilliant appearance when in blossom; it clothes the lower part of the tree, and becomes a tree itself, killing the one which assisted it to climb. The colour of the flower varies from scarlet to yellow, and is often quite spherical in shape. It resembles the Golden Bottlebrush of Australia, though brighter in colour. It flowers in Autumn, and the other Ratas in Spring.

DENDROBIUM CUNNINGHAMII.

Native Name :—HIRI-TURITI.

A VERY large tropical Asiatic genus, of which one only is found in New Zealand. D. Cunninghamii is a tufted epiphyte with masses of cylindrical white roots and slender stems growing on trunks of trees. Throughout the Islands as far south as Stewart's Island. *Hook. fil.* It grows near the top of Rangitoto and also at the Great Barrier Islands, Auckland Harbour. It flowers in December.

DENDROBIUM CUNNINGHAMII.

PLATE 17.

METROSIDEROS ALBIFLORA, *Banks and Sol.*

White Rata.

THIS is probably the handsomest of the climbing species of *Metrosideros*, or Rata. The leaves are large, broad, and long-pointed, and the flowers, which are white, are arranged in large leafless cymes at the ends of the branches. It is confined to the northern parts of the North Island, and may often be seen clothing the lower portions of the trunk of the Kauri Pine, or others of the larger forest trees. (*Cheeseman*). Flowers in November. Thames.

METROSIDEROS ALBIFLORA

FUCHSIA PROCUMBENS.

Native Name :—TOTERA.

—·· ～～◁◈▷～ ··—

THIS fine genus abounds in parts of South America, but has hitherto been found in no other country except New Zealand, which is one of the most remarkable features in the distribution of the genus on the one hand, and of the New Zealand flora on the other. One kind forms a large bush or tree (F. excorticata) and the other a small prostrate plant. *Hook. fil.* It grows in the sand near the sea, where it was found by Professor Kirk and received the name of Kirkii, but it has since been united with F. procumbens. It has the peculiarity of the flowers growing upright; the blue anthers are also peculiar. It flowers in December. Auckland.

ALSENOSMIA MACROPHYLLA.

Native Name :—PÉRE.

—·· ～◁◈▷～ ··—

ALSENOSMIA is a genus confined to New Zealand and to the north of the Northern Island. It was discovered by Sir Joseph Banks, and afterwards by Allan Cunningham; the name he adopted was suggested by the powerful fragrance of the flowers, J. D. Hooker. The name is well deserved—the scent is delicious, and is retained for a long time by the flowers. It is a shrub 6 to 10 ft. high. It flowers in Spring. Waikomiti, Auckland.

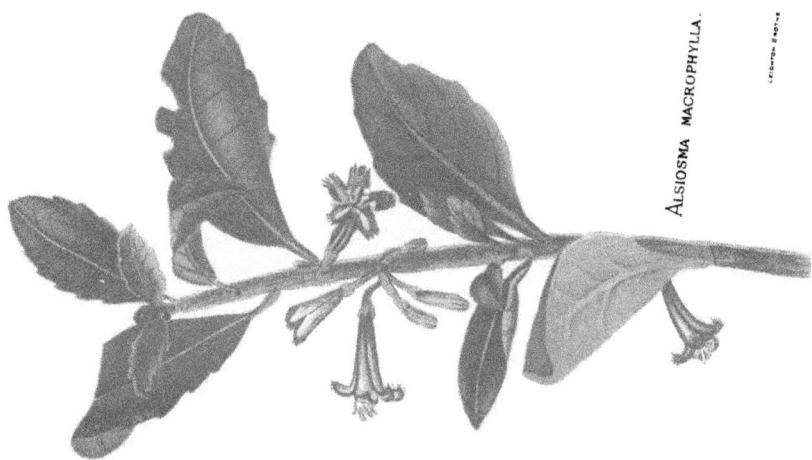

ALSIOSMA MACROPHYLLA.

FUCHSIA PROCUMBENS.

PLATE 19.

SENECIO HECTORI, *Buch.*

A BRANCHED, woody shrub, 6 to 12 ft. high. This remarkable addition to the flora of New Zealand was collected by Dr. Hector on the Buller River, Nelson Province, 1872. The magnificent floral display of this species and others, such as S. glastifolius, with similar white rayed flowers, can only be seen to advantage in their natural humid habitats. It also grows between the river Mangles and the Tuangahua, and inland from Collinwood. (*Buch*). It flowers in January. These two Senecios were drawn and kindly contributed by Mr. John Buchanan, Geological Survey of New Zealand, by whom they have been already figured in Trans. New Zealand Institute.

SENECIO ROBUSTA, *Buch.*

A SMALL woody shrub; branches covered with scales, formed by the sheathing cases of the old petioles. Allied to S. monroi and S. longifolia, but differing very much from both in habit of growth, very coriaceus leaves, with peculiar venation, and small robust coryrubs of few large heads of flowers. Trans. New Zealand Institute. Also drawn and contributed by Mr. John Buchanan, Geological Survey of New Zealand.

[20]

Senecio Hectori. Buch.

Senecio robusta. Buch.

PLATE 20.

OLEARIA ILICIFOLIA, *Hook. fil.*

A SMALL shrubby tree, not uncommon on the mountains of both the North and South Islands. Both flowers and foliage have a pleasant aromatic fragrance when fresh. The leaves have a peculiar serrated and crinkled edge, hence its name. It flowers in January. Mount Egmont.

CELMISIA GLANDULOSA, *Hook. fil.*

A SMALL alpine plant growing on Mount Egmont and at the base of Tongariro and Mount Cook. The leaves are of a bright glossy green. It flowers about December.

FORSTERA BIDWILLII, *Hook. fil.*

THE Genus Forstera belongs to a large and very interesting natural order (Stylidiaceæ), which has its headquarters in Australia. Forstera itself embraces but four species, of which one is confined to the alps of Tasmania, and the others, which are possibly only varieties of one, to New Zealand. F. Bidwillii is a native of the Northern Island ; it is found on the ranges of Mount Egmont, on Tongariro, and the summit of the Ruahine range, in shady places. (*Nicholson*). It flowers in January.

CELMISIA LONGIFOLIA, *Cass.*

A VERY widely distributed plant, found both in the lowlands and in mountain districts. The genus Celmisia may be said to be one of the chief ornaments of the herbaceous portion of the New Zealand Flora, containing nearly 30 species, nearly all of which possess considerable beauty. (*Cheeseman*). Flowers in December. Arthur's Pass.

[21]

Celmisia longifolia.

Celmisia glandulosa.

Olearia ilicifolia.

Forstera Bidwillii.

PIMELEA PROSTRATA, *Hook. fil.*

Native Name :—TAUHINAU.

————~•~————

THIS pretty little shrub must be familiar to most people in New Zealand, especially in Auckland, where it grows in great abundance ; the long sprays of masses of little flowers make it look like large white feathers, but the colour soon changes to a light brown. It grows amongst the tea-tree shrub (Leptospermum Scoparium) on clay hills. "It is a member of a very large Australian and Tasmanian genus. They abound on all the coasts, especially of New Zealand, forming small shrubs, easily recognised by their decussate leaves and very tough bark used for cordage, paper, etc." *Hook. fil.* Flowers in November.

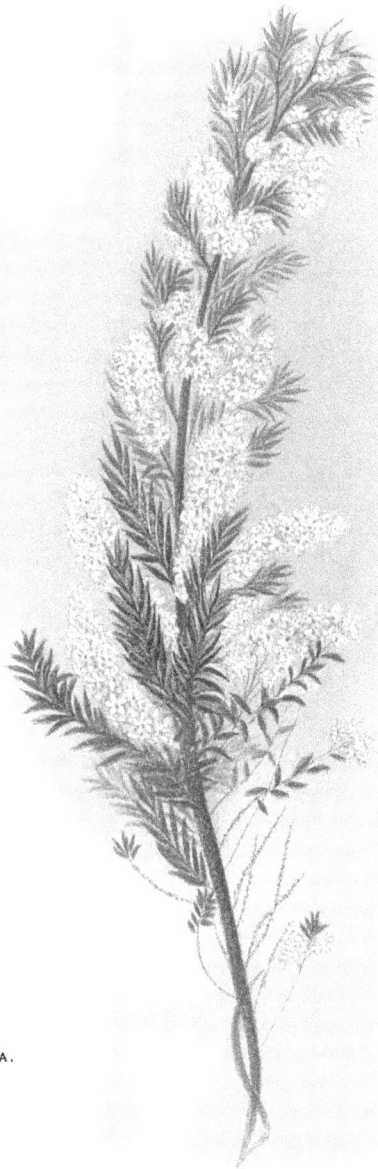

PIMELEA PROSTRATA.

PLATE 22.

LIBERTIA IXIOIDES, *Sprengel.*

———

NOT uncommon throughout the whole of the Islands. Growing in damp, swampy places. Leaves sometimes stiff, rigid, and of a bright yellow colour, and also grassy, as in the plate. Flowers very tender and fugacious, of the purest white. (*Cheeseman*). It flowers in November and December. Thames.

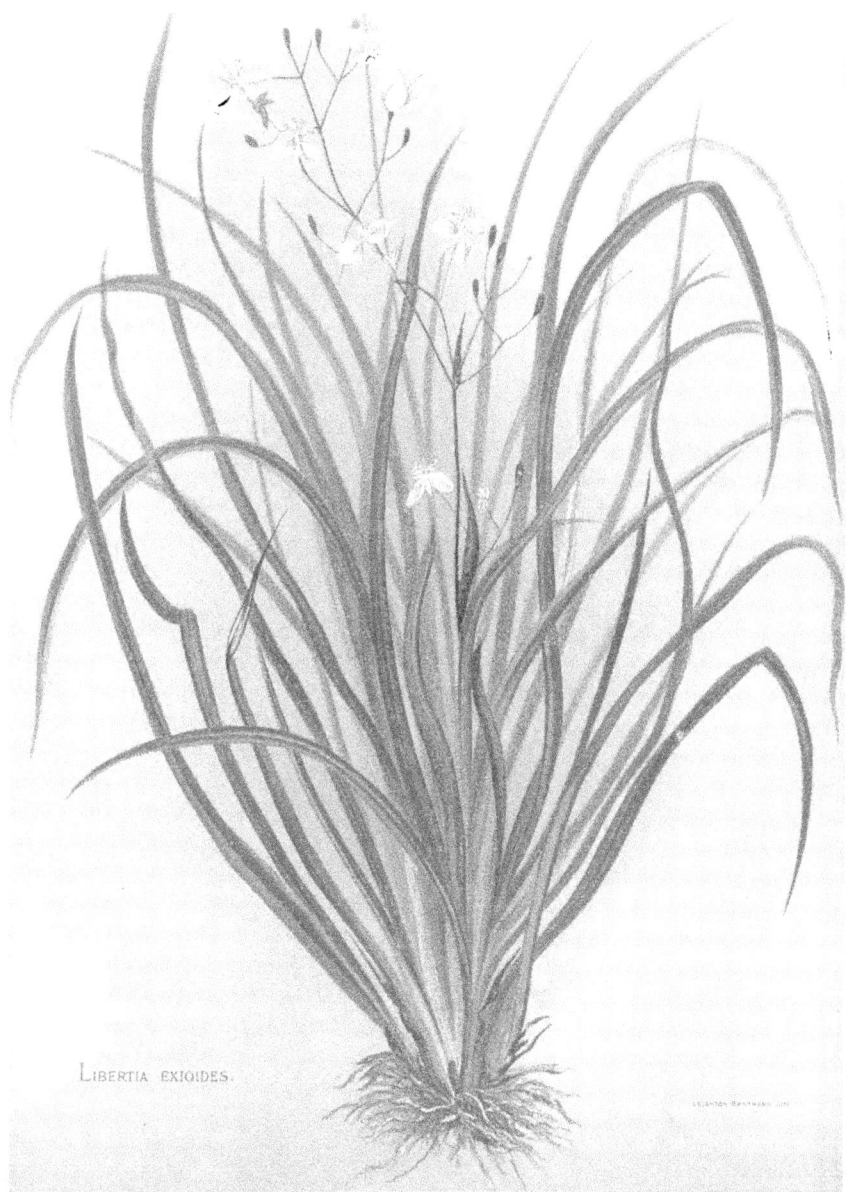

LIBERTIA EXIOIDES.

PLATE 23.

GENTIANA SAXOSA, *Forst.*

Common Gentian.

———◦•◦———

An abundant sub-alpine and alpine plant throughout the whole of the mountain districts of New Zealand, attaining its greatest luxuriance in open grassy flats at elevations of about 3,000 ft. The large white flowers are decidedly handsome, but cannot compare in beauty with the more brilliantly coloured blossoms of some of the European and American species. (*Cheeseman*). Flowers in February. Lake Rotoiki.

———————————

LINUM MONOGYNUM, *Hook. fil.*

White Flax.

Native Name :— RAUHUIA.

———◦•◦———

This delicate little plant, with its pure white flowers and pale green leaves, grows principally about Christchurch and the Canterbury province. It is the most beautiful of its tribe and is easy of cultivation. It differs from the English flax plant in being perennial. The flowers vary in size from one-fourth to nearly an inch across. It blossoms in England in the greenhouses in May and June, having a succession of flowers for a great length of time. *Hook. Ic. Pl.* It flowers in February.

GENTIANA SAXOSA.

LINUM MONOGYNUM.

PLATE 24.

MERYTA SINCLAIRII.

Cabbage Tree.

Native Name.—PUKA.

———◦———

ONE specimen alone of this fine genus of singular looking trees has been found in New Zealand—two others inhabit Norfolk Island, and a third Tahiti. They form erect, slender, small trees, with long simple trunks. *Hook. fil.* The specimen represented came from the Hen and Chickens (small islands at the entrance to Whungarei Harbour), which, I believe, is its only habitat. It has been transplanted to Auckland, where it grows freely in some gardens. Flowers in December.

MERYTA SINCLAIRII.

ARISTOTELIA RACEMOSA, *Hook. fil.*

New Zealand *Ribes ?*

Native Name :—MAKO-MAKO.

⁓⁓⁓⁓

THIS is a very beautiful shrub or small tree, which springs up on the edge of clearings and at the side of roads after the trees in the forest are felled. It is curious that it is not seen before, but as soon as light and air are let in, it covers the ground. It is found in great quantities throughout the Islands, but especially at the side of the coach roads in the South. The head of flowers of all shades from crimson to white and the pink under the leaves give it a very handsome appearance. Flowers from October to January. Wai-wera.

ARISTOTELIA RACEMOSA.

CALCEOLARIA SINCLAIRII, *Hook. fil.*

ALL the species of *Calceolaria* are confined to South America, with the exception of two native to New Zealand—the present one and *C. repens*—which thus afford a good illustration of the affinity existing between the floras of the two countries. C. Sinclairii is, however, of very local distribution, being confined to the district between the East Cape and Napier. The flowers, though pretty, are very small compared with those of many of the American species. (*Cheeseman*). It flowers in March.

EUPHRASIA MONROI, *Hook. fil.*

Eyebright.

THERE are probably only about a score of members of this genus, which really deserve specific rank, although a large number have been described as distinct. The Eyebrights are found in temperate regions of both hemispheres, about five being represented in the flora of New Zealand, and these are distinct enough from those which occur in Arctic Europe, &c. The generic name is derived from the Greek *euphraino* to delight, the plants being supposed to possess the power of curing blindness. In rustic practice in Britain E. officinalis is still used as an eye-medicine. (*Nicholson*). It flowers in January. Mount Egmont.

[27]

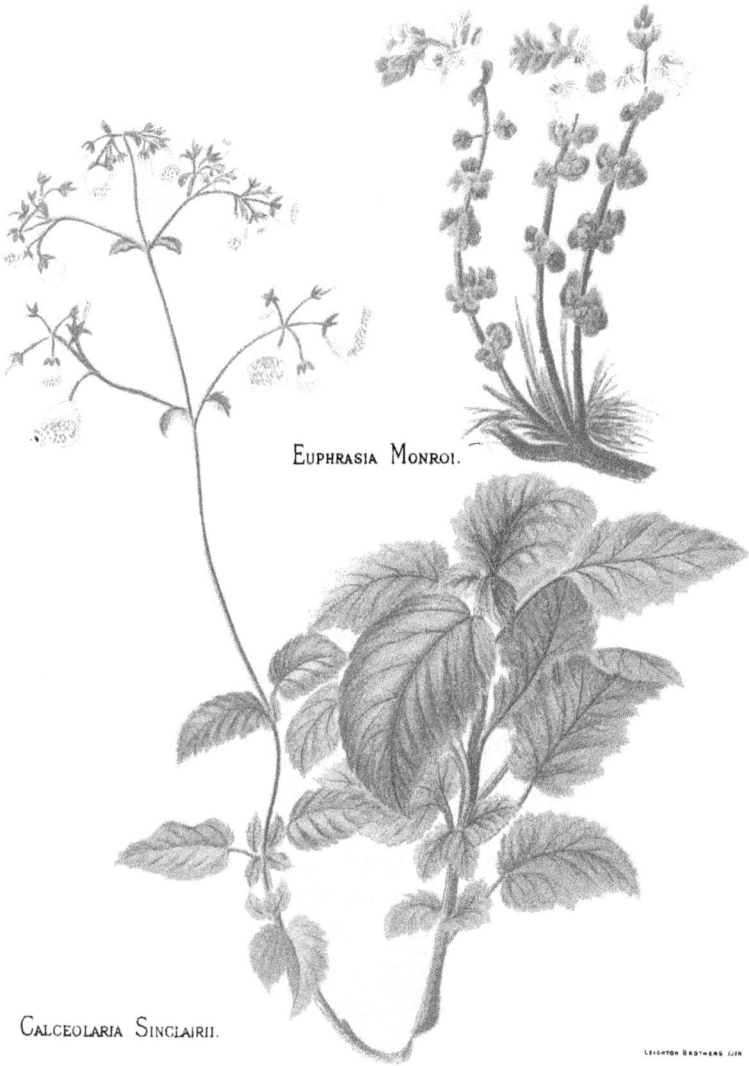

EUPHRASIA MONROI.

CALCEOLARIA SINCLAIRII.

POHOTUKAWA.

METROSIDEROS TOMENTOSA.

PLAGIANTHUS LYALLII, *Hook. fil.*

Lace-Bark Tree.

Native Name :—WHAU-WHAU.

A SMALL tree, 15 to 25 ft. high, common in the mountain forests of the South Island, especially by the sides of streams. It has a graceful, spreading mode of growth, fine bold foliage, and exceedingly handsome large white flowers, and is altogether one of the most handsome and noteworthy trees found in the Colony. The bark is remarkably tough, and is often used by the bushman as a substitute for rope. (*Cheeseman*). It flowers in January. Otira Gerge.

Olearia insignis.

www.ingramcontent.com/pod-product-compliance
Lightning Source LLC
Chambersburg PA
CBHW021816190326
41518CB00007B/624